I0487526

THE SCIENTIST'S DOG

Copyright © 2013 and 2018 Lawrence T. Friedhoff
All rights reserved.

No part of this publication may be copied, stored in an archival or other similar system, or transmitted by any means, including but not limited to: electronic, mechanical, photocopying, recording, scanning, or otherwise without the prior written permission of the Copyright holder.

Limit of Liability/Disclaimer of Warranty: The author and publisher make no representations or warranties with respect to the accuracy or completeness of the contents of this book and specifically disclaim any implied warranties of merchantability or fitness for a particular purpose. No warranty may be created or extended by sales representatives or written sales materials. The advice and strategies contained herein may not be suitable for your situation and should not be implemented without guidance by your own professional advisors. You should consult with a professional before taking any action based on the information in this book. Neither the author nor publisher shall be liable for any loss of profit or any other commercial damages, including but not limited to special, incidental, consequential, or other damages.

Cover design by Lawrence and Sarah Friedhoff

ISBN: 1500513997
ISBN-13: 978-1500513993
Pharmaceutical Special Projects Group, LLC (PSPG) Publishing,
New Jersey, USA

To order additional copies visit www.thescientistsdog.com

TABLE OF CONTENTS

Warning

Do not try this at home. Consuming beverages concocted with random food-related substances is not recommended by me or anyone else associated with this little book.

Prelude

This story begins on a freezing cold winter night in New Jersey. I had spent the day outside in the snow and the cold.

Me, out in the snow

My work for the day was done and a quiet night of reading by the fire seemed both appropriate and desirable. A hot drink would make an excellent addition to a good book.

Tea seemed like a good idea. We had lots of different kinds already. But, I am a drink chef. I like to invent new beverages. The chai smelled interesting. It had cinnamon, which I always like in tea, especially if it has that special spicy "bite" found in some varieties. Not being one to leave well enough alone, I improvised: some dried herbs from the garden, a pinch of mushroom powder from a friend who markets nutraceuticals. Finally, a little bit of cheap Scotch never hurt anyone, and it mixes well with tea, especially if milk is also added (a secondary benefit is that the cinnamon and milk cover up the "chemical" taste of the alcohol in the Scotch). Nice, very smooth and pleasant to drink. Perfect for a late evening in front of the fireplace in the depths of winter.

The tea was more powerful than I expected. Did I really add that much Scotch? Which herbs did I use? I wasn't sure exactly what I had added. No matter, the slight dizziness might make it a bit harder to concentrate on my new book, but I knew I'd probably fall asleep before I'd read more than a few pages.

Maybe that's what happened. But... maybe it was something unexpected.

A quiet night by the fire

Chapter 1 - An Unexpected Comment

Based on its title, the book seemed like it might be interesting; at least to me; an analysis of a special method of stock-market investing based on demographic trends and sociologic analysis of political movements. I knew I would hate the mumbo jumbo but, I had to manage my own pension, and maybe I'd get some good investment ideas. Over the years, I've found that good ideas often come from unexpected sources.

The book began with a rather dry discussion of free markets: each seller trying to get the highest possible price, and each buyer trying to buy at the lowest possible price. "God, it sounds so exhausting," I mumbled under my breath; "much simpler to go to the store and pay the price they ask for most things. Why bother with an auction for my next pair of shoelaces."

"I agree," said Henry. "In fact, why bother with shoe laces at all. They aren't really necessary. I can't see why you even bother with shoes for that matter."

Sorry, I didn't mention Henry earlier. He had been sitting at my feet in front of the fire. Since he's a dog, it didn't occur to me to mention him.

Henry had never spoken to me in words before

Henry had never spoken to me in words before. Of course, he would whine for food or jump and growl when he had to go out. But that was the extent of it. Nothing out of the ordinary and nothing like full sentences strung together.

The amazing thing was that it didn't seem strange that my dog was speaking to me in just the kind of voice I would have imagined Henry would have: measured and thoughtful, but a little lower pitched than one would expect for a small dog, sort of like a kindly professor.

In retrospect, I suppose it might all have been just a dream. On the other hand, as you will see when you read on, it was very unusual for a dream. Maybe it was the tea.

Chapter 2 - A Different Perspective

"Well you're very talkative tonight!"

"Well, you're very talkative tonight. Why the sudden interest in auctions and shoelaces, when you haven't said a single word in the entire time you've lived with us?" I asked.

Henry looked me straight in the eyes in a way he never had before. "I guess, I've just had enough. I know, I'm the dog and you're the Master. You say jump and I don't even get to say 'how high?' I just have to jump. But, I've had enough. The things you're doing are just too dumb for me to remain 'dumb'. I had to speak out. If only for my own mental health."

Silence hung between us. I must admit I was a bit taken aback. I've been criticized from many directions, but, a man expects his dog to be loyal. "You're the last one I'd expect to give me a hard time," I said, trying to use my least petulant voice. After all, he was my dog, and I didn't want to let him think he could get the better of me.

"Is it criticism for a seeing-eye dog to pull a blind man out of traffic? That's what I'm doing. You're so obviously blind and wandering dangerously. I had to do something. You should be thanking me." Henry sounded a little bit hurt.

"What in God's name are you talking about? I'm not wandering anywhere. I'm sitting in front of a nice, warm fire trying to enjoy this book. I think I *was* enjoying it, at least a little, until you started talking. I must admit, I really don't know how to talk to a dog."

Henry just shook his head in obvious disbelief. "Right, you don't know how to talk to a dog. Do you know how many times I've had to hear you discuss my bladder and bowel movements when we're out in the yard? 'Do business, do business!' and 'hurry up, hurry up!' It's cold, I want to go back in the house!' You seem to know perfectly well how to talk to me. Yammering on and on, talking about things in your selfish, self-centered way. My goodness, excretion is something that shouldn't be discussed in polite company. You know perfectly well how to talk to me. You just don't know how to *listen* to me."

He was right. I didn't know how to listen to him. But, apparently, I was going to learn. Would I have to put up with suppressed criticism accumulated over the five the years since he came to live with us? It was all very confusing, and I guess in retrospect, it's amazing that I wasn't completely flabbergasted at the whole experience. But, at the time, it seemed almost normal, if a little irritating. Of course, I wasn't expecting what came next.

Chapter 3 – A Dog's Eye View

"You can't help it." Henry said. "You're basically blind, just like all humans. It's amazing that your species survives at all. We dogs would like to help. You people are so cute, even though somewhat 'out of it'. Wandering around, worrying about the most silly things while totally oblivious to the important things." Henry said this in a somewhat wistful way. As if he wished it weren't so, but knew he was powerless to change things much.

Henry said this in a somewhat wistful way. As if he wished it wasn't so but knew he was powerless to change things much.

"I don't know if there's much hope," he said. "I've thought about doing this for a long time and always believed it was hopeless. But, I'll try. Maybe I'm too pessimistic."

I considered myself a bit more intelligent than the average dog, and so I was a little insulted by his attitude. I was determined to show him who had the superior intellect.

"I was speaking about auctions and shoelaces. Apparently, you have an objection to both. But I assure you that both are very important. They are concepts beyond the understanding of a dog!" I said.

Henry shook his head and started out gravely. "There's a difference between understanding and acceptance. I understand what an auction is. I just know it's stupid. And as for shoelaces, I suppose if one is stuck with hairless, padless, poorly-developed feet, shoes and shoelaces may be a necessity. But, that's just evidence of how poorly adapted humans are to this place we all live. You are all weak and confused. But, I suppose that's what makes you so cute and endearing most of the time.

"But, things are changing," he continued, "and you humans seem completely oblivious to what's going on. How can one teach the blind to see?" His brow was furrowed in a way that made it seem like he was really considering the question.

"I know!", he exclaimed. "I know just how to put it in simple terms even you can understand."

Chapter 4 – Professor Dog

That's why it's where all of beauty resides.
In the small things, the incomprehensible and uncategorizable.

Henry began to explain. "Life is full of fleeting things: a breeze, a flash of lightning, the sound of a sneeze off in the distance. These come and go and we barely notice them. In fact, most of what happens in the world are things like these: numerous, hard to categorize, and fleeting. That's what the world is, mostly. Like the brilliantly colored leaves in the Fall, suddenly picked up by the wind and scattered here and there. Some propelled high in the sky and some scuttling along in the gutter. It's all beyond comprehension, let alone understanding. That's why it's where all of beauty resides - in the small things, the incomprehensible and uncategorizable.

"You humans try to capture this in art; an admirable but sadly pathetic exercise, always trying to 'understand' or 'express your artistic sense.'

Can't you see that the world is not 'understandable'? Not even with those stupid machines with the odd screens, and funny symbols, and keyboards. No, life isn't 'understandable' on that level. If only we could get you humans to see this, perhaps, we could save you from your fate, which is so obvious."

"So, my first lesson for you is, forget about understanding the world or anything that's really important. It's simply not possible. Anything that can be well understood, is, of necessity, unimportant."

I thought about it. In a certain sense, I suppose he was right. We all like to make simplifications when we think about things. Einstein used what he called "thought experiments" to help him understand our very complex world. In his thought experiments, Einstein constructed an imaginary world that was much simpler than the real world we live in. For example, he said he conceived one of his most famous "thought experiments" as a child, when he imagined riding on a light wave; thinking about this possibility and what it would be like gave Einstein some of his earliest insights into the special theory of relativity. But, of course, it's not possible to ride a light wave in the real world, that's what makes it a "thought experiment" rather than a real experiment.

Well, all of this may have been beside the point. Henry was asserting that "understanding" the physical world was a hopeless exercise. This was a direct challenge to my occupation, my life's work. As a scientist, my goal was to understand the world based on the scientific method, and the results of real-world experiments. I was proud to be part of a long tradition stretching back beyond recorded history: Archimedes, Galen, Newton, Curie, Einstein. Who was Henry to tell me it was all for naught. For that matter, it clearly wasn't for naught. Without science, we wouldn't have modern versions of agriculture, sanitation, medication, or communication, just to mention a few of the ways science has improved our lives.

"Henry, you're just saying that understanding isn't important because dogs can't do it. You're justifying your own species' inadequacies by saying they're unimportant."

He shook his head in mild disappointment. "I knew you'd say that. Humans are so predictable. Arrogant, self-centered, and, of course completely ignorant when it comes to knowing anything important."

"I will try to dumb this down for you. Nothing *important* is understandable. Only simple, unimportant things are understandable, even for beings with grotesquely over-developed frontal lobes on their brains."

"I'm not saying that nothing is understandable, just nothing that's *really* important. I'm not even saying that trying to understand things is unimportant. You humans clearly enjoy trying. You're constantly yammering on about it. Debating the Keynes versus Friedman's theories of economics. Making mathematical models of social security payments 50 years in the future. Trying to find the *cause* of climate change. These are important issues, but they're simply not understandable. So, all this talk is just that, *talk*. It's a form of recreation, or worse, a form of deception that you humans seem to enjoy in a way other species find completely mystifying."

"So, as step one, I'd like you to give up on the idea that it is possible to 'understand' economics, climate change, Alzheimer's disease, et cetera." He spoke imploringly, as though he hoped I would understand but didn't expect me to.

The last one hit home. Having led the team that developed the main treatment for Alzheimer's disease and having searched for newer treatments on and off for years, it would be hard for me to accept the idea the cause of Alzheimer's would remain beyond reach forever.

Henry could see that I was finding it hard to accept what he was saying. So, he followed up right away.

"Let me give you an example that I think will help you," he said with a little more optimism in his voice. "You're a physician, and I'll wager you think you know the *cause* of pneumococcal pneumonia, right?"

For those of you who aren't familiar with it, pneumococcal pneumonia is a relatively common infection of the lung; its cause is well known – a tiny, round, infectious bacterium called the pneumococcus that grows in pairs. Pneumonia caused by pneumococcus used to be a common cause of death, but since the mid-20th century, it has been curable with antibiotics

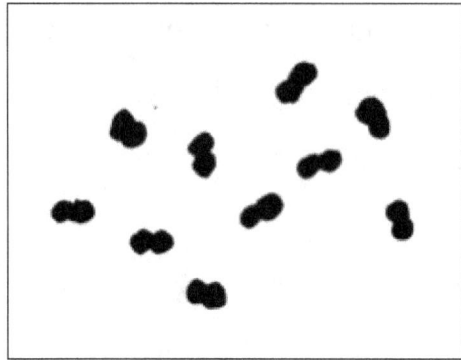

Pneumococci, the cause of pneumococcal pneumonia

"Yes, I do think I know the *cause* of pneumococcal pneumonia." I said emphatically. "I've treated patients with pneumococcal pneumonia and

I was able to cure them *because* I knew what caused the disease. The fact that I was able to cure them shows that I *understood* the cause. I UNDERSTOOD it! Right?"

"Wrong!" he exclaimed. "Wrong, wrong, wrong! You knew how to help your patients, but not because you understood what caused their disease. If the pneumococcus is the cause of pneumococcal pneumonia, how do you account for the fact that pneumococcus is spread all over our environment. There's probably some growing on your skin right now. "The patients you treated for pneumococcal pneumonia probably had pneumococci on their bodies for years, yet they didn't have pneumonia except on the one occasion you treated them. You don't know why they got pneumonia when they did, or why the billions of other people carrying pneumococcus haven't gotten pneumococcal pneumonia and probably never will. Perhaps, one patient gets pneumococcal pneumonia because he is made more susceptible by a prior influenza infection, perhaps another is made more susceptible by poor nutrition, or by exposure to cold weather, or some or all of these things combined, or because of a million other things that we will never identify. These are just as much the "causes" of pneumococcal pneumonias as the pneumococcus bacteria. In that sense, we can never really 'understand' this illness nor can you identify a unique cause. As I said, before, 'nothing important is understandable', nothing."

"But, that doesn't mean nothing is understandable." He went on. "Simple things are understandable. You speak and people hear. That's understandable. You mix a bunch of hydrogen and oxygen in a rocket engine and light it, it explodes and pushes the rocket forward. That's understandable." You ignite some TNT in a sealed container near some people, and it explodes blasting off their limbs and killing them. That too, is understandable. But, the greater consequences of your speech, the result of space exploration, the aftermath of a war: these are unpredictable and not amenable to understanding."

I was beginning to get the idea and I had to admit that it didn't seem quite as unreasonable as I had expected when we started out. What was unreasonable was that it was coming from a dog.

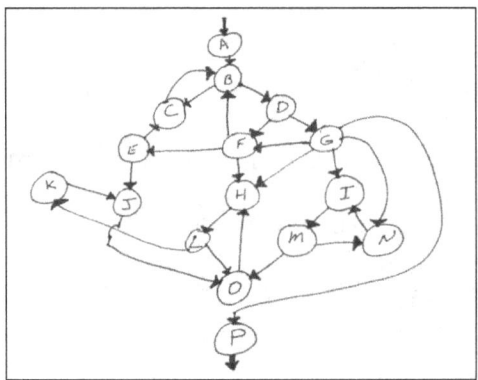

Complicated? Not compared to the real world!

The Earth as seen from Apollo 17 (NASA)

Chapter 5 – Unimportant Things

"You see," Henry continued, "humans excel at understanding unimportant things, transient things like farming, plumbing, electricity. You believe that because these things make life a little easier, they are *important*. But, over the long run, they aren't. They only make things easier over the short run.

"Yes, humans have increased in numbers exponentially based on the understanding that led to these *advancements* in technology, and yes, things have gotten somewhat easier for some humans because of these advancements. But the changes are temporary on the time scale that we dogs understand. Eventually, the advancements lead to other problems: population growth, global warming, environmental degradation. You know about many of them. As these complications of technology accumulate, you humans will end up right back where you started from or be even worse off."

"After all, the technological advancements not only lead to complications like global warming and over population, they include the development of advanced land mines, cluster bombs, nerve gas and death camps... ugh, there are too many terrible consequences to list. Just thinking about them is hard for a dog. Communicating with you about them is even worse. We dogs can't conceive of why a species would develop in the way you humans have. You've even perverted the wonderful life-giving power of the sun to make horrible nuclear weapons."

I could see Henry was deeply troubled… That it was painful for him to think about these things. I could see that these thoughts had weighed on him for some time. I guessed he had waited to discuss them because these thoughts were so painful.

"So, you're saying that all of human scientific and technical progress is worthless? That I've wasted all of my professional life trying to help sick people by understanding their illnesses and devising new treatments?"

Everyone knows that attacking a person's sense of self worth isn't likely to make them sympathetic to a new point of view. I wondered why Henry was taking this tack with me. I was trying not to feel threatened, but it was hard.

Henry shook his head. "I'm not saying it's worthless, just unimportant over the long run compared to many other things that you humans are oblivious to. Playing Frisbee isn't worthless, it's just unimportant over the long run. Most of life's activities in themselves aren't important, but that doesn't make them completely worthless… just unimportant over the long run.   This is especially true because by ignoring the important things, you make the unimportant things even less important."

"Let me give you an example," he said, appearing a little more hopeful that I might understand. "Let's say you are walking along tossing a ball up and down on a beautiful day. You toss the ball up and catch it. Each time you catch it you feel a little sense of success. It's a simple thing, catching a ball. So, the sense of success is also little, but it's real and pleasant."

I thought, "So far, so good, he's making sense."

Henry continued, "Now, let's pull back a bit from this happy scene, and we see that as you are walking along enjoying your catching your ball, you are heading toward a huge cliff. You're concentrating on the ball, having a nice time, but not for long. The fall off that cliff will undo the pleasure of the ball catching with a vengeance. Does that mean that the pleasure of playing with the ball is worthless? No, just unimportant compared with other things that are going on at the same time. Does that make sense to you?" He looked at me hopefully…

I said, "Sure, it makes sense. I just wish I knew what cliff you were talking about." I supposed he had some doomsday prediction in mind. They've come and gone for millennia, "the end of the world is near… blah, blah, blah…" But the end never seemed to arrive. This would make Henry just another failed prophet; just another nudnik. I felt a lot better, and Henry's challenges to my self worth started to fade.

But Henry kept at it. "I keep telling you, the important things in life aren't understandable in the way you suppose. I'm not predicting the end of the world, I'm saying it isn't predictable. The end of the world is important! That's why it can't be *understood,*" he said gravely.

"I wouldn't begin to try to predict the end of the world or anything else like it. But, that doesn't mean we can't do something about it. Just like treating pneumococcal pneumonia, even though you don't really understand the cause. Do you get that?" Henry was asking a real question, but I didn't have a clue as to how to answer, or for that matter what he was *really* asking.

"I'll explain how we can do something about things that we don't really understand, in a minute," he said. I could see he was getting tired.

Chapter 6 - Important Things

Henry's eyes were closed and his brow was furrowed. He was clearly thinking about how to put things in terms that I could understand.

"You see, understanding is grossly over-rated. You don't really understand how a car works yet you are happy to drive it. You know enough about it to start it, drive it to where you want to go, get out, lock it and so on. You don't really understand what makes rain. But you know that it comes from time to time, and that you need a roof on the house to keep it out. You also know that it will likely come often enough to make your garden grow. So, you don't *understand* how a car works, or what makes it rain on one day rather than another. But that kind of understanding is unnecessary, and in some real sense, unobtainable. Now, these aren't really important things. But they are good examples of situations that you don't need to understand in detail to know what to do."

"Important things are always like that. Though they aren't understandable or analyzable in detail, we can still know what to do and how to try to avoid problems."

"But the first step is to recognize that these important things aren't amenable to a detailed understanding or accurate prediction.

Acceptance of this fact makes everything that follows easy. We dogs have accepted this for millennia. That's why we've been so successful. We live for free, get free medical care, have no work, and no real stress. We enjoy the simple things: running, eating, smelling the fresh morning air, relationships with others. We have no wars, our murder rate is almost non-existent, we have no crime or taxes. We have created an almost perfect utopia and you self-centered, arrogant, ignorant humans could learn a lot from us."

"But, of course you humans are too busy fighting about money to bother about anything important. And, that preoccupation would be fine for us, except that your insanity is beginning to threaten all the species on the planet. It is threatening the stable and fulfilling way of life we dogs have maintained for millennia. We've done our part, and we have been enjoying the fruits of our labors for longer than we can count. But now, you humans may wreck everything because you aren't paying attention to the most important things."

"Okay, so what are the important things we humans are ignoring and that are so threatening to you?" I asked. I was a little annoyed that I was being accused after providing Henry with, what he admitted, was a very comfortable lifestyle.

"Well, let's start at the beginning," he started. "Stone tools were fine. They made things a bit easier and they were completely recyclable. No chance of running out of stone; most of the planet is made out of stone, and there's not enough energy coming from the sun to allow changing it very much. So, that part was okay."

Fire, by and large was okay too. As long as you kept to burning leaves and wooden logs, it wasn't a problem; just shortcutting the carbon cycle a bit, depriving some microorganisms of food. But the microorganisms would have converted most of the leaves and logs into carbon dioxide and water, just like burning does, anyway. So, that wasn't too bad, at least for us dogs."

"Metal, well, that's where you started to go wrong. First it was metal that exists in nature, like copper, that can be found pure, rather than extracted from ores. Sure, metal made better tools than stone. But, you humans must have realized that pure metals weren't being formed as quickly as you were using them - those rare concentrations of pure copper were quickly being used and dispersed and wouldn't re-form for a very, very, very long time. You must have known that at some point, the metal that was easily findable would be used up. So, I guess you humans had a plan for what you would do when that happened, right?

WRONG! No plan… just a hope that you would find some way to get by."

"Now, you see, we don't really know where the copper metal found in nature comes from, or why it forms in the kinds of veins or nuggets that we can find in specific places. But, we do know that it isn't forming very rapidly, and that we are likely to deplete it if we use it more quickly than it is being formed. So, although we don't know the details, or the exact time when it will become very difficult to find enough new pure copper, we do know copper will start to run out at some time. And that's all we really need to know."

"Of course, humans got around the problem of running out of readily available pure copper by discovering iron and steel and inventing ways of extracting copper from ores that contain it in a less useful, oxidized state. This delayed the copper problem, but didn't make it go away, for all the reasons we know without having to understand."

"Metals are a limited resource. They exist only rarely in nature. To get them in quantity, they must be extracted from sources where they exist in a less usable, more dispersed form. To do this requires work, and work requires fuel, and we know that fuel is also limited."

"Originally, work was all done by muscle power and the 'fuel' was just food. The energy in the food was derived from the energy of the sun, so it was, for all intents and purposes, unlimited. But, over time fuel became non-renewable. Coal, lignite, natural gas, oil and all the smelly rest: these 'fossil' fuels supply their energy by using up the oxygen in the atmosphere.

"OXYGEN!!! We can't live for more than a few minutes without it! The most critical element supporting almost all life on Earth, animals AND plants. And you humans have the brilliant idea of using it up! Depleting a critical resource that we all have shared for millennia in the cycle of life and photosynthesis. You humans decide to start taking more oxygen out of the atmosphere than is going in."

"Now, we don't need a detailed understanding of the dynamics of the atmosphere, or a computer model of climate change, or even an understanding of why our lovely planet was endowed with a beautiful atmosphere containing so much oxygen, to know that if more oxygen is taken out than is put in, we are likely to have a long-term problem. Does that make sense to you? Take more out than is going in, and over time there will be less and less. You don't even need numbers to understand that: no measurements to high precision, no computer models. Take more out than is going in and eventually, you will run out."

"And, we know we'll have problems long before we 'run out' because we need a lot of oxygen to live. When the levels get too much lower than what we have now, we'll get sick. We know we can expect this from what happens at high altitudes where there's less oxygen."

"Above about 30,000 feet, life as we know it, is impossible. Each year, as you humans deplete the atmosphere of oxygen, we all essentially move up the mountain and there's a little less oxygen. Now that's pretty important, wouldn't you say? Running out of oxygen!"

"But, it's worse. You see, oxygen is being replaced by carbon dioxide. More carbon dioxide is going into the atmosphere than is being taken out. So, even without computer models, analyses by Nobel laureates, and the like, we know there will be more and more carbon dioxide in the atmosphere over time. In fact, anyone who has ever been trapped in a confined space knows that the air slowly gets 'stale' with increasing levels of carbon dioxide and falling levels of oxygen."

"The increased levels of carbon dioxide are unhealthy too. They make us work to breathe harder and make our kidneys work harder to get rid of the excess. And, by the way, this extra work requires that we metabolize more food using up more oxygen, and generating even more carbon dioxide. Not a lot more, but more. This is what we dogs call a 'positive feedback loop' or a 'self-reinforcing' process."

"Self-reinforcing processes are very important, over the long run because they grow over time. In fact, in the long run, self-reinforcing processes are the ONLY things that are important. And you humans have set up a bunch of new ones that threaten all species, including yours."

"So, you see, it's the self-reinforcing processes that are important. These are often recursive processes. As time goes by, the progress of the process affects its future progress, over and over and over again. Because of this, even a small change in the way the process affects its future behavior can have very large effects on how fast it goes and the stability of its progress. That's because with each cycle, the small changes compound. This feature of many self-reinforcing, compounding processes is what makes them impossible to understand in detail and makes it impossible to predict their behavior, in detail."

"We can never measure the properties of each individual cycle well enough to know what will happen after it compounds a hundred or a thousand or a million times. In fact, when processes feed on themselves like this, over and over, even quantum effects can become important, and quantum effects have an inherent element of unpredictability. You know about quantum mechanics, right? The Heisenberg Uncertainty Principle?" Henry was asking a real question. I could see it in his eyes. He was really unsure if I understood the Uncertainty Principle of quantum mechanics."

"I know about the Heisenberg Principle," I replied, feeling a little annoyed that a dog was lecturing me about science. "And, I guess it makes sense that the evolution of recursive processes could be hard or even impossible to predict. But, you're saying that all the important things in life are governed by recursive processes. Is that really true?"

"It is over the long term." Henry replied, clearly relieved that he didn't have to explain quantum mechanics to me. "If you understand the mathematical formalism of quantum mechanics, you can easily see how pretty much everything is governed by recursive processes—the Hamiltonian operator causing systems to evolve over time; recursively operating on the state vector over and over again."

"But, a simpler way to think about it is that essentially nothing in the world is static. Everything changes, at least over some time frame. And that change is clearly dependent on where things are before the change."

"Mountains grow or erode away at a rate that is dependent on how high they are. Rivers change their course in a way that's dependent on what it was to begin with. This repeats itself over and over. Actually, rivers are a good example, because often we can't predict how the course of a river will change over long periods of time. The outside of the curve of a river erodes more quickly than the inside, causing bends to become bigger and bigger. So, a bend that may be too small to notice can grow very large over time. On the other hand, at some point the bend becomes too extreme and the river suddenly straightens out. All this is unpredictable in detail."

"Okay," I admitted. "You've convinced me that <u>many</u> important things in real life are not understandable in detail. Maybe you're right that <u>all</u> the important things in life are not understandable in detail. But, if I accept that, how can I justify my life as a scientist. I am trying to understand things in detail, and I am trying to predict how changing something will affect a complex system like a human being. For example, I try to understand a disease and then design a treatment to improve its symptoms or cure it."

Although I had accepted the basis for Henry's argument, I rejected the conclusion that scientific analysis had no long-term importance. I didn't think I was rejecting it because it was an implicit rejection of my life's work; a rejection of the importance of what I did every day. It was more a gut feeling. Based on a lifetime of observation, I knew that science was relevant. That it could be used to develop new technologies that were significant advances and that these advances made the world a better, happier, healthier place. I had seen this with my own eyes: the world was a better, happier, healthier place now than when I was a child, and much of that improvement was, as far as I could see, based on scientific progress.

"I'll accept the idea that maybe everything important in life isn't understandable in detail." I relented. "But science does play an important role in life in our modern world. I'm sure of that. So, explain how those two concepts fit together—why detailed scientific understanding of the real world seems to be so important, and yet, the important things aren't amenable to detailed scientific understanding? Explain that!" I demanded. How do those two things fit together?" I asked. The insult to my ego at being instructed by a dog now seemed less important than my scientific curiosity. I genuinely wondered about how these two apparently contradictory ideas could be reconciled, since I now believed both of them.

Chapter 7 – Reconciliation

"We'll, it's not hard to reconcile the two. They really aren't contradictory at all." Henry began. "Simple systems can be understood in detail and their evolution over time can be predicted accurately. Complex systems, like those making up almost all of the real world, can't be understood in detail nor can their progress be predicted in detail. But, that doesn't mean complex systems can't be understood at all, nor does it mean that we can't know how to deal with their uncertain evolution over time.

"Let's take the simple example of a river flowing downhill over a relatively flat plain that I mentioned earlier. Any little variation in the bed of the river can divert the flow enough to start the river to change course a little, so that it's no longer completely straight. Once the path of the river curves a little, the outside of the curve erodes a little more than the inside, and so, the curve gets bigger.

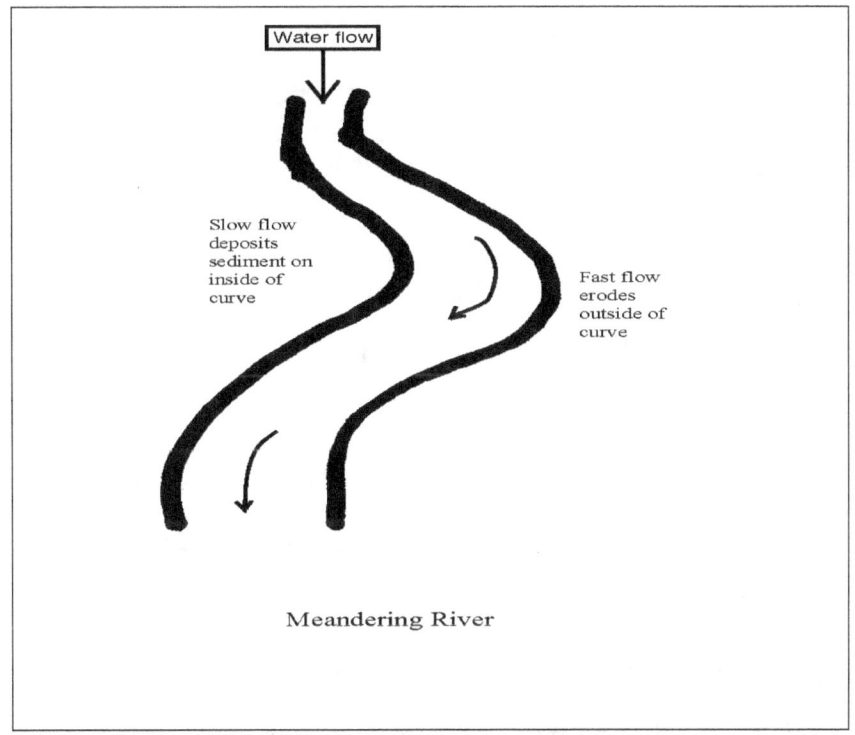

Picture of a Meandering River – It really happens this way!
(Source: US Fish and Wildlife Service)

"As it gets bigger, the outside erodes more quickly still, and the curve gets bigger still. And so on, and so on, until there is a big curve and the river pushes through it and straightens out again."

"We can't know exactly where the curves will be or when they will appear. But, we do know they will appear, grow, and eventually disappear in an endless cycle. We know this as certainly as we know anything, because we understand the basic physical laws underlying the process. We can't know how the river will evolve exactly. BUT, we do know that it will, and we can make some very good predictions about what the extent of the changes in the course of the river will be. So, we can say that it's probably unsafe to build a house in a certain place, because the river will always be moving, and it could move to that place. Conversely, we know about how far a river can deviate from a straight line downhill, so we can identify places where it is very unlikely to meander."

"It's just like physical chemistry; you don't need to know the detailed movements of every molecule of a gas to know what its pressure will be, and for most purposes, knowing the pressure is all that's important. Knowing the pressure tells you how strong a tank you need to contain the gas. You don't need to *understand* the gas in detail to know what its pressure will be, and that pressure may be all you need to know to buy the correct tank. Of course, it is conceivable that for some reason, you really need to know where a particular gas molecule will be three weeks from now. In that case, you're really out of luck. Its exact position is completely unpredictable." Henry looked at me hopefully, thinking I would probably both understand and accept this.

Being a former physical chemist, this was something I could easily grasp. Statistical mechanics tells us that if we understand the basic laws governing parts of a complex system, we can often predict its average behavior in ways that are very useful, if not complete in every detail. This kind of statistical analysis is key to the rapid technological progress we've experienced since the 19th century, when the field of statistical mechanics was first invented.

"Sure, that makes sense," I said. "We don't have to understand every detail about how a complex system will evolve over time to know what to do to deal with the many different ways it can evolve."

"Exactly!" Henry replied. "The real world requires a different kind of analysis, based on the prediction of general properties and an understanding of how to deal with the many possible outcomes that can occur. You don't need to know where every molecule of a gas will be to know what kind of tank you need to hold the gas. Figuring out the strength of the tank is far, far simpler than predicting the state of the gas in detail. Simple enough so that a dog can understand it and do it."

I couldn't miss Henry's sarcasm. He clearly thought that his species was far superior to humans when it came to understanding the world. On the other hand, he was trying to explain things to me, so he must have thought I had enough intelligence to grasp what he was trying to express, otherwise, why would he bother? It seemed likely that it was a great effort for him to try to communicate with me. In a way, I was flattered that he would take the time and effort. I was beginning to feel like I might actually learn something from him.

This set my competitive juices flowing. I've always prided myself as being better than average at learning from others. It's kind of like turning a profit, of sorts, on every conversation. I've always felt that the sign of real intelligence was the ability to learn from others. Intelligent people can learn from those who are less intelligent more than vice versa. We all have good or even brilliant ideas. The more intelligent among us are just more likely to recognize them as such. It seemed to me that Henry, in spite of his species, might have some really interesting ideas, and I was now anxious to absorb them. I guess, in my subconscious, I thought I might be able to appropriate them as my own. After all, who would ever believe they came from a dog?

Chapter 8 – The Stench

"You see, it often isn't necessary to understand the causes of things, or to have a detailed mathematical model of their behavior in order to understand what to do about them. This is lucky, because as we've now agreed, most problems aren't amenable to a detailed 'understanding.' You don't need a degree in hydrodynamics to water the lawn or to use an umbrella. One can know what to do about a problem without 'understanding' it in detail." This was Henry's way of resolving the apparent contradiction we had been discussing.

"I guess I have to agree." I said. "In the real world, science only <u>helps us</u> identify problems and their solutions; and that's really important. But, we can never understand the real world in the scientific sense. It's just too complicated. And furthermore, since we are part of the real world, we would have to understand ourselves in order to understand the real world, and we humans accept the idea that we are not completely analyzable from the scientific point of view."

"Well, that's just your human arrogance. You think you are so complicated that you could never be subject to complete scientific analysis. But, it's a good start. If you can just extrapolate that to the important things in the world as a whole, you will begin to understand the way we dogs look at things," he said, looking a bit relieved.

I realized that dogs, with their leisurely and privileged way of life, probably weren't used to intellectual conflict, and that our conversation might be very tiring to Henry. He could see the sympathetic look on my face and he said… "You're right. This kind of conversation is a bit tiring for me. However, these things must be said."

"I'm flattered that you feel the need to tell me all these things. I do feel like I've learned a lot from you" I said, feeling that sense of superiority I get when I learn something that I think is more than what was intended to be communicated; attributing the increment to my inherently greater intelligence.

"We're not done. Not by a long shot." Henry replied, the energy in his voice growing stronger. I have some important things I need to tell you, but we'll have to go over some basic science before I can really explain them to you. I'll try to make it easy so you can understand. But first, I want you to take a deep breath through your nose and tell me what you smell."

I was suspicious because everyone knows dogs have a better sense of smell than humans; a lot better. But, how could I refuse such a simple request?

I took a breath through my nose and said..."Nothing, I don't smell anything."

"Try again," he said. "Take some short sniffs, like you were a dog trying really hard to pick up a faint smell. You've seen me do it a million times. Do it like I do."

Now I felt a little silly. I was being asked to mimic a dog, and I knew he could always tell me I wasn't doing it right because he was a better judge of how dogs sniff than I was. Nonetheless, I tried it. I took a few short sniffs, but for the life of me, I couldn't smell anything.

"Okay, I've tried, but I still don't smell anything. You can tell me what I'm missing." I said, somewhat defensively.

Henry was trying to look non-judgmental. I could see it in his eyes. But, I could also see that he found me woefully inadequate.

"Understandable. You humans are cursed with a woefully inadequate sense of smell housed in a stunted, withered nose. It's part of why you miss so many important things. You see, most of the important things in life are characterized more by smell, the sense we dogs excel at, than by sight, the sense you humans rely on so much. Sight gives too much detail; too much data. More than can be reasonably interpreted. The detail obscures the important things. Smell, on the other hand gives an average. The unimportant details, the things that are transient and fleeting are washed away, and the important things are easier to spot. If you humans relied on smell more than sight, you'd have a much easier time of it. But, that's why I'm here. I'll explain part of what you are missing."

"You see, the odor of the world is changing. The air is becoming more acid, more irritating. You humans can't smell it, but for us dogs it's unmistakable. It's the smell of all the burning you humans are doing: coal, oil, natural gas, lignite, ethanol, diesel. Your catalytic converters hide some of the smell, but the residue is unmistakable to a species like ours that has a fully-developed sense of smell. It's the carbon dioxide accumulating in the atmosphere. There's no doubt about it to us dogs. We've smelled it for the last hundred years or so, and it's been getting worse and worse. Of course, the carbon dioxide is replacing precious oxygen, and we dogs can smell that too; the small but noticeable decline in oxygen is very scary to a dog. We know how critical oxygen is to almost all life on earth."

"But, that's not all. There are other smells that go along with it. The smells of a warming earth; of melting glaciers, of melting tundra, of forest fires… all of these are combining to a sickening *stench*… a *stench* that you humans seem totally oblivious to. And they're all related."

"As you humans burn more fossil fuels, the air gets thicker with carbon dioxide. The carbon dioxide holds more heat, which melts more tundra, which in turn releases more carbon dioxide, which holds more heat. It's a self-reinforcing cycle. Thus, it is inherently unpredictable. It may go quickly, or it may go slowly, but it IS going, and there is no particular reason to expect it to stop and lots of reasons to believe it will accelerate."

This is all common knowledge to us dogs, so we rarely talk about it. But, it's really what I want to talk to you about. I want to explain one of the ways you humans have gone so very wrong, and to explain what you have to do about it. Give me a minute to collect my thoughts and energy."

"Since what I want to explain is important, it isn't amenable to complete or detailed understanding, nor, can we predict in a detailed way how it will change with time. But, I think you will agree that it's important, even if you are oblivious to the horrible *stench*. And I think you will agree we have to do something about it."

I gazed into the fire waiting for Henry to regain his energy. I wondered what would come next.

Chapter 9 – Starting from the Beginning

Henry began again. "On a clear night, far from city lights, gaze up into the sky, and you will see a small part of the vast universe that surrounds us. With a small telescope or binoculars you can see further and with the most powerful observing instruments scientists can see almost 14-billion light years into the vastness. Fourteen billion light years, with each of those light years being almost one trillion kilometers (over 600 billion miles)."

"The size of this expanse is really beyond imagination. It is so much larger than any space here on Earth, that its extent defies comprehension. Yet, throughout this monumentally vast expanse, this stupendously gigantic extent, we have never observed more than a trace of something we have here on Earth in vast quantities. Something so common, we almost never think about how precious it is to us. Something that we must have every second of every day of our lives, and without which we would all perish in a few minutes. This substance is molecular oxygen."

"Molecular oxygen makes up 21% of our atmosphere. So common around us that we take it for granted, and almost never notice it. Yet other than on the surface of the Earth, molecular oxygen at these at these concentrations has never been observed anywhere else in the universe.

On the scale of the Universe, it may be one of the rarest substances, and yet, we never marvel at our luck in having it in such abundance."

"Almost none of us really knows where it came from or why it is all around us, and more importantly, almost none of us ever contemplates the possibility that it may not be here forever. That one day, Earth could become much like the rest of our observable Universe: a place with no meaningful molecular oxygen; a place without the kinds of life that we take for granted; a place without people, birds, frogs, horses, trees, flowers or lions. You never think that our Earth could become a place inhabited solely by putrid sulfur-metabolizing bacteria and other similar creatures that now are confined to small regions of our planet by the molecular oxygen we crave but which is lethal to them. Humans never think that one day our planet could be their planet."

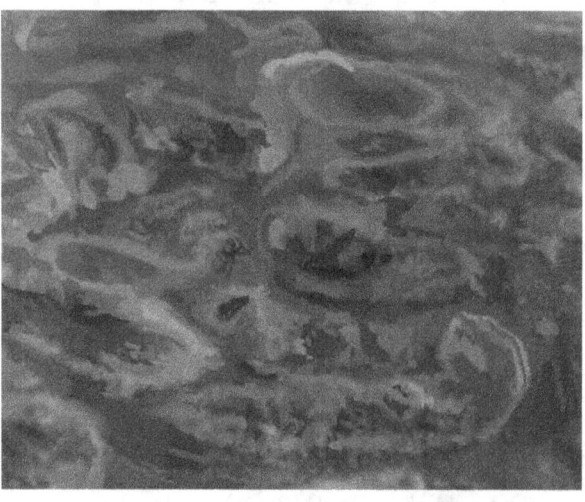

Green and purple sulfur-eating bacteria. These bacteria will likely thrive in a low oxygen world.

"We dogs remember tales of the Permian-Triassic transition, a time on the Earth when the oxygen concentration of the atmosphere and oceans dropped far below their levels today. During this period, which lasted several million years, almost all the species on Earth perished, most likely as a result of the lack of molecular oxygen in the atmosphere" (See *Additional Reading* for sources of information about the Permian-Triassic).

"All the life forms we see in the world around us today are descended from the few species that survived that horrible, choking time when the breathable part of our Earth's atmosphere may have almost vanished entirely. We can assume that had the oxygen concentration declined just a little more than it did, the species alive today would be very different, and perhaps, might not have included us. Yet we know virtually nothing about what caused the decline in atmospheric oxygen, and therefore, have no idea of what could cause it to happen again. What we do know for sure, is that the amount of oxygen in our atmosphere is extremely rare in that part of the Universe we can see, and that therefore the conditions that give rise to the oxygen in our atmosphere may be remarkably rare and, on the timescale of the Universe, very fleeting."

"I'm going to remind you of the importance of molecular oxygen. I'll explain the processes that tend to consume it, and what could cause oxygen consumption to rapidly outstrip oxygen production, giving us another Permian-Triassic transition or worse. Before doing this, I will embark on a brief explanation of some of the scientific concepts that will be needed."

"I'm going to put it in simple terms, so that you can explain it to other people, people who maybe aren't as familiar with science as you are. Remember how I've explained it, and pass it on. It's the only hope we dogs have that you humans may change your ways and save us all from a terrible catastrophe."

"Terrible catastrophe" seemed a bit extreme to me. I found it hard to believe that a dog could conceive of a global catastrophe or a global anything, for that matter. But, I was interested to hear what came next.

Chapter 10 – What Henry Said

I can't remember every word that Henry said next. But a lot of it was about things I already knew, and I can summarize for you here. He had a remarkable grasp of basic scientific principles and how they related to one another, and he put them together in a way I found pleasing and elegant. In retrospect, I wondered if he understood what was going on the day he got a static electric charge from the pool cover. Given what he explained to me, I'll bet he did. Though of course, I didn't think so at the time.

Henry all charged up with static electricity from the pool cover

I'm not going to give you a complete course in thermodynamics. However, thermodynamics plays a crucial role in what Henry was trying to tell me. In fact, thermodynamics plays a crucial role in pretty much everything that is important to humans; therefore, a basic understanding of thermodynamics is critical to a modern person's thinking about the world we live in. Our understanding of thermodynamics, gained during the last part of the 19th and the early part of the 20th century, made the Industrial Revolution and our modern lifestyles possible.

Thermodynamics is based on several "laws" or principles. Henry was concerned with two of these, the "first" and "second" laws of thermodynamics.

The first law of thermodynamics states that the energy of a closed system, though it may change its form, is always conserved. That is, the total amount of energy in a "closed" system cannot change. To some degree, this is just a matter of definition, because when the amount of energy in closed system is observed to change, we define a new kind of energy that keeps the total the same, or else we define a kind of energy that is being transferred into or out of the system in a way that keeps the total the same.

As it turns out, energy and mass are equivalent (according Relativity Theory), so it may be possible to measure the absolute amount of energy in a system by measuring its gravitation. However, this is not practical in most situations, so we can think of energy as a quantity that cannot be created or destroyed but can change form as long as we are willing to keep defining new kinds of energy in a way that keeps the total constant.

Here's a simple example. Consider a ball sitting still on top of a hill. Gravity will tend to pull the ball down the hill. We can define the ball's energy as its mass multiplied by both the height of the hill (the ball's elevation) and the force of gravity. If the ball rolls down the hill without friction, the energy related to its former position at the top of the hill is reduced, since the elevation is reduced. However, we know by definition that the total energy must remain the same, so we have to define a new kind of energy related to the ball's motion; if we define it properly, we find that the sum of the energy related to elevation plus the energy related to motion will always total the same value, i.e. the total energy is conserved.

In reality, no ball can roll down a hill without friction, so in the real world, the energy related to the ball's height plus the energy related to the ball's motion will never be quite constant. This requires that we define yet another form of energy relating to frictional forces. If we define this friction-related energy properly, then we will again find that the total energy remains constant.

Physicists have gone to great efforts to define new kinds of energy so that the total will remain the same. These include gravitational energy, energy related to electrical and magnetic fields, energy related to pressure, energy related to nuclear forces, and energy related to quantum effects. When defined and measured properly, the sum of all these energy values remains the same no matter how a system changes.

One kind of energy is, however, different from all the others—heat energy. Heat energy is sometimes difficult to define, however, it has special properties that make it a powerful force in the Universe, because heat energy is related to the extent of order in a physical system. Because of the unique nature of heat energy, it tends to always increase as a proportion of the total amount of energy in a system. This tendency of heat to increase results in the tendency of all systems toward a more homogeneous state and is called the second law of thermodynamics. As a general rule, systems will not become completely homogeneous because the first law of thermodynamics, which limits changes of total energy, will prohibit certain arrangements that would alter the total energy.

Thus, the ultimate state that any system, including presumably the Universe itself, tends toward, is a balance between the states allowed by the first law of thermodynamics and the tendency of systems to become more and more homogeneous as determined by the second law. The actual state of the world and everything in it is a result of a kind of struggle between the first and second laws of thermodynamics, and knowing that both these laws are, so far as we can tell, never violated, they provide a way to help us understand complex systems. This is because systems like weather in the atmosphere, human physiology, rivers flowing downhill, can never, ever, violate these laws. So, they provide some constraints on how anything in the world can change over time, even very complex systems. The Second law of Thermodynamics will be the subject of the next section.

Chapter 11 – Henry Explains The Second Law of Thermodynamics

Henry continued his exposition, more or less as follows.

The second law of thermodynamics is one of the foundations of modern physics. In its simplest form, this law states that in the real world, things that we can distinguish from one another tend toward greater and greater homogeneity. For example, mountains and valleys eventually erode and fill in, so that without some other balancing process, their height differences become less and less over time.

The second law can also be seen in everyday activities. For example, put water into a bowl of dry dog food and the water will gradually spread through the dry dog food so that it is no longer concentrated on the bottom of the bowl but becomes more or less evenly distributed. Although in accord with the first law of thermodynamics, the density of water at the bottom of the bowl is very slightly higher than at the density at top, due to the force of gravity, which tends to pull the water molecules more toward the bottom of the bowl. The ultimate distribution of water is determined by a balance between the first and second laws of thermodynamics.

The same considerations apply to our atmosphere. The Earth's gravity tends to pull all the air down toward the surface of the earth. The second law tends to have it spread evenly over the entire Universe. The compromise between the two laws is denser air near the surface of the Earth and thinner air high above. We experience this compromise between the two laws when we climb a mountain or fly in an airplane.

This tendency for differences to dissipate towards homogeneity as driven by the second law, within the constraints of conservation of energy as required by the first law is the basis for essentially all things that are important to human beings. For example, staying clean: when you take a shower, you bring a large volume of clean water in contact with the dirt on your skin. The second law tells us that there will be a tendency for the dirt to spread out into the water, and for the water to spread on your dry skin. This allows you to clean most of the dirt off your body at the cost of getting wet. You then bring a dry towel into contact with your wet skin, and the second law again mandates that the water will tend to leave your skin drier and make the towel wetter. These are just a few of many examples of the second law that can be observed in everyday life and that are vital to your survival.

The second law is important in another way. In many situations, the "tendency" of things to "even out" can be very powerful. For example, the "tendency" of the high temperature of burning gasoline to want to cool off to the temperature of the world around it (thus dissipating the temperature difference) is what provides the energy to turn a car's wheels. In fact, essentially all your engines, generators, nuclear reactors, batteries, etc. produce the energy used in everyday life by capturing the force generated when "differences" of some kind are dissipating. You use the forces generated when these differences are dissipated to reverse the effects of the second law in small locations.

Thus, the dissipation of temperature differences in a gasoline generator can be used to make electricity to run an air conditioner. The air conditioner can pump heat from a cool place (the inside of an air-conditioned house) to a hot place (the hotter outdoors). So, the temperature difference between the inside and the outside of the house increases. This is not a violation of the second law of thermodynamics, because the amount of dissipation of heat difference occurring in the electric generator will always be larger than the increase in heat difference created by the air conditioner.

Net differences in temperature are always decreasing overall but may increase in small areas. This "local" reversal of the second law is what allows essentially everything that is important to you. It allows you to dry the wet towels and to purify the water you use for your showers.

Every action you take that results in less homogeneity, e.g. cleaning the basement, raking leaves, organizing your desktop, must be balanced by greater homogeneity elsewhere.

In the case of the human body, the transformation of food and oxygen from separate substances into more homogeneous ones with less contained 'chemical energy' (carbon dioxide, water, and feces) is what allows our survival. Without dissipation of the separation oxygen from the ingredients in food that is mediated by your metabolism, human life could not exist.

Such "separations" are so important, that a unit of measurement has been defined to measure their magnitudes. This unit has been given the name, 'free energy.' Free energy can actually be quantified in many situations and is part of the total energy needed to keep energy conserved. When free energy can be defined, its numerical value is a measure of the amount of work that can be extracted from the "separation"; e.g. a measure of the amount of electricity that could be generated, concrete that could be hoisted, or water that could be pumped. The remainder of the energy, the "non-free" part, is not available to do useful work.

Because, the second law mandates that the "differences" that contribute to free energy cannot increase but rather must dissipate over time, free energy will always tend to decrease.

Since the Universe is quite old (around 14 billion years, according to recent estimates) shouldn't all the differences have dissipated by now? Shouldn't the free energy of the Universe be zero, you might ask? As far as we can tell, the second law holds for the Universe as it now exists. So, free energy is indeed decreasing with every moment that passes. The free energy that still exists is the result of an incomprehensible and inexplicable event, the explosive creation of the Universe we call the "Big Bang."

This creation event was an enormous violation of both the first and second Laws of thermodynamics: all the free energy of the Universe was created in a single instant—all the free energy that powers the shining stars and galaxies, all the energy encompassed in the mass of the Universe (because energy and mass are equivalent, $E=mc^2$). Since the creation of the Universe, the total free energy has been steadily decreasing, and some day, in the very far distant future, it will approach zero.

It is impossible to know for sure whether the first or second law will eventually win out: if the first law wins, the Universe may eventually end up as a pinpoint of super-dense material merged together in one powerful black hole; if the second law wins, then the Universe will end up as a cold, homogeneous bore. At the moment, it seems like the second law is ahead, but of course, we will never know for sure, and it is possible that neither law will win out but rather there will be some compromise final destiny. But the long-term is too distant to concern anyone but cosmologists and theologians. We live on a much, much shorter time scale.

In the meantime, you use the tendency of free energy to decrease to run every element of your society and world. You help the second law along by dissipating inhomogeneities and extracting part of the free energy to do work that is useful to you: from the energy metabolism of your bodies, to the engines that power cars, ships and planes to the bombs that you people use to kill one another. Without this free energy, there could be no life, no purposeful movement, and no change.

For this reason, free energy is the most important thing we can have. I want to discuss some of its intricacies, particularly our most important source of free energy, the separation of oxygen from hydrogen and carbon that serves as the power source of almost all life on earth.

Chapter 12 – Where Does Earth's Free Energy Come From?

As Henry explained earlier, the evolution of our Universe, including our planet, is the result of a kind of struggle between the first and second laws of thermodynamics. The second law mandates that things even out, especially temperature, so that the entire Universe tends to become homogeneous. The first law mandates that only certain arrangements that conserve energy are allowed to exist. Within the constraints of these two basic laws, we find the concept of free energy to be a kind of mediator; it is the energy related to the remaining inhomogeneities that haven't yet succumbed to the second law, such as temperature differences that haven't yet equalized, rocks on top of mountains that haven't yet rolled down, etc.

Free energy is very important because it can be changed into heat energy and in the process do what we call "work." For dogs, work is the power to run across a field or bite something really hard. For humans, work may be moving a car uphill, doing the laundry, taking a breath, or flipping a switch. Free energy can make things happen that would never have occurred by themselves. As this free energy is dissipated into heat, part of it is reincarnated as something we want, perhaps, a clean kitchen. The dissipation of free energy is the process that allows all happiness.

Since free energy results from "separations," it is found in many forms. The Earth and Moon attract each other by their very powerful gravitational forces. The separation of these two bodies, which is maintained by the Moon's orbital movement around the Earth, gives rise to ocean tides, differences in the height of the oceans' water. In this way, the free energy stored in the separation of the Earth and Moon is gradually decreasing as it is transferred to the height differences in the Earth's oceans, which in turn dissipate through friction that converts the differences to heat energy.

Similarly, nuclear reactions inside the Earth that stem from the partial resolutions of inhomogeneities of subatomic particles give rise to temperature differences between different parts of the Earth's interior. These differences in temperature are a source of free energy that is sometimes used to do work through geothermal-generating stations.

However, the largest source of free energy for our planet is the free energy from the Sun's radiation. This energy is released primarily as the result of separate hydrogen atoms merging together into helium atoms. This is the first step in a much longer process that results in the merging of all parts of atoms into homogeneous kind of nuclear matter that is slowly occurring throughout the Universe. The amount of free energy this first step generates in the Sun is almost beyond human comprehension (though Henry claimed that dogs can understand it).

A tiny part of this free energy arrives as sunlight on the Earth's surface. Some of the free energy of this sunlight is absorbed by the Earth's surface and converted to heat energy that is then re-radiated back into space. This radiation sent back into space has a lower free-energy content than the sunlight it came from, as the second law requires.

A tiny portion of the free energy in the sunlight that falls on the Earth's surface is captured by plants. They use this free energy to make their tissues grow through the process of photosynthesis. This free energy is used to separate carbon and hydrogen from oxygen, thereby allowing plants to carry on vital synthetic and metabolic processes that give rise to their limbs, leaves, roots and other parts. In the course of carrying out these vital processes, plants store some of the sunlight's free energy by separating carbon and hydrogen from oxygen. With the exception of the tides and some heating of the earth from interior nuclear reactions, essentially all the free energy available on Earth comes from sunlight. Part of this sunlight free energy is captured by the process of photosynthesis used by plants directly to make their physical structures. In this process some of the free energy is also stored by plant's separation of carbon and hydrogen from oxygen. This stored free energy is available to both plants (that made it in the first place) and animals, almost all of which absolutely require it for survival and reproduction.

Everyone knows that food (a source of hydrogen and carbon) is required for life. Even more acutely necessary is an adequate source of oxygen: life ends quickly without it. This is because recombination of hydrogen and carbon (in food) with oxygen (from the air) is essentially the only source of free energy available to maintain life in animals. Without it, there could be no heartbeat, no muscle movement, no conscious thought, and no happiness.

Because oxygen is so prevalent, we tend to take its availability for granted. It seems to be a constant in a changing world. However, it is clear that the level of oxygen in the earth's atmosphere has changed considerably over the history of our planet, sometimes being considerably higher than it is now, and sometimes plummeting to levels that are incompatible with most life (including canine and human life).

As would be expected from the critical function of oxygen for living things, changes in atmospheric oxygen have sometimes had a profound impact on the Earth. Therefore, it is prudent to consider oxygen in more detail: where it comes from, where it goes, and how levels in the atmosphere may change. Of particular importance are the large decreases in atmospheric oxygen in the past have that seem to have led to mass extinctions and which, if they occurred again, could lead to the extinction of all dogs (and humans).

Chapter 13 - All The Chemistry You Need to Know in One Chapter

Henry explained that dogs had understood chemistry for far longer than people "You see, we dogs are chemically oriented. Our sense of smell is far superior to that of humans. So, we think in chemical terms, because we are sensing chemical changes in the air all the time. I'm sure you can understand that knowledge gained by superior senses is often far more important than intelligence, as you humans measure it."

"Without sensation, in this case the ability to perceive very important changes in the atmosphere, you humans are blind to very important changes that occur around you all the time. Sure, you can measure these changes with scientific instruments. But, numbers changing on a dial don't carry the same meaning as sensing things directly, as we dogs do, with our superior sense of smell. That's why we've understood the important concepts of chemistry for far longer than humans."

Henry couldn't fully explain chemistry as dogs think of it, but he tried to express what he could in terms a human like me could understand.

"The universe is full of matter, and almost all matter (and essentially ALL matter on our planet) is comprised of atoms, which are small agglomerations of even smaller particles. Atoms are important because most are stable for very long times and can be rearranged to form all of the different substances and chemicals of our everyday life."

"Although the equations that govern the structure of atoms are very well understood, the computations necessary to predict the structure of most atoms is far beyond the capability of the most powerful computers. For this reason, chemists have developed some simple models that can predict the way atoms behave most of the time. These over-simplified rules are correct most, but not all, of the time. The study of these imperfect rules is called 'Chemistry.' It is far from an exact science—it is more like a collection of 'rules of thumb."

"Since essentially all matter on Earth is comprised of atoms, it is easy to understand that the simplest substances, from the chemist's point of view, are those comprised of a single kind of atom. Such substances are called 'elements.' We are all familiar with some of the elements; hydrogen, (the most abundant element in the universe) that blew up the Hindenburg Zeppelin; helium (second most abundant in the universe) that we use to inflate party balloons; oxygen (that we need to breathe); uranium (that we can use to annihilate our species), etcetera."

"As it turns out, atoms can combine with one another in various ways to form new substances. These new substances are called 'compounds' because they contain more than one type of atom. These combinations of atoms are also sometimes called 'molecules.' As might be imagined, the mathematics required to predict the ways atoms combine is much more complicated than the mathematics required to understand a single atom. Thus, solving the exact equations to predict how atoms combine is far, far beyond the capability of even the most powerful computers. Nonetheless, chemists have derived simplified methods, or rules of thumb, which can often (but not always) predict how atoms will combine with one another."

"One of these rules of thumb is that when two or more elements are put in contact with one another, they tend to mingle and combine. In this combination process, one kind of atom can share part of its components with another. In a way, this is the result of the second law of thermodynamics, which tells us that in time the elements will not remain separate but will commingle to make a more homogeneous substance."

"One set of commingling atoms is of particular importance—the combination of oxygen atoms with the atoms of other elements. As it turns out, oxygen is a particularly sociable element, tending to combine with many other elements if it should encounter them. Much of the oxygen on earth is combined with the element silicon. This combination is called silicon dioxide (dioxide because there are two atoms of oxygen for each atom of silicon). Silicon dioxide is the main component of quartz and glass, and makes up a significant part of the surface of the earth. Oxygen also binds to hydrogen to make water. A water molecule contains two hydrogen atoms and one oxygen atom. On other planets, much of the oxygen is combined with carbon to form carbon dioxide (again "di" because there are two oxygen atoms for each carbon)."

"Because oxygen is so promiscuous, it is rarely found in its pure form (a gas, comprised of particles that have only two oxygen atoms stuck together and do not include other kinds of atoms). This form of oxygen, the gas that is found in our atmosphere, is called 'molecular oxygen,' because it is not comprised of single atoms of oxygen, but rather of pairs of oxygen atoms held together by shared parts. Molecular oxygen is incredibly rare in the Universe and therefore probably represents our most precious resource."

Chapter 14 - Where Does Molecular Oxygen Come From?

Molecular Oxygen is Rare

As Henry said, high concentrations of molecular oxygen, such as those found in the atmosphere of the Earth, are remarkably rare. In spite of decades of searching, molecular oxygen has not been found in quantity on any other planet in our solar system. Small amounts have been found in the atmosphere of Saturn's moon, Rhea, but these amounts are likely far smaller than what is found on Earth.

Furthermore, searches of the known Universe have not turned up any other evidence of meaningful concentrations of molecular oxygen. Admittedly, molecular oxygen becomes increasingly difficult to detect as celestial objects become more distant; however, among the thousands of planets found outside our solar system to date, no molecular oxygen has been observed. This suggests that molecular oxygen is very rare in our Universe and that therefore, life, as we know it is also very rare outside of our Earth.

The high concentrations of oxygen in the Earth's atmosphere are likely due to a very rare combination of circumstances. Since these circumstances seem to be quite uncommon on the scale of the known Universe, we should not presume that they will persist indefinitely even here on Earth. This is especially true, because we know from fossil evidence that high concentrations of molecular oxygen didn't always exist here. Given the incredible rarity of molecular oxygen and its critical importance to life on Earth, we should, Henry urged, probably consider molecular oxygen to be the most precious commodity in the Universe. It may be one of the few substances essentially unique to our planet.

The Earth is a "complex system," and thus, not amenable to detailed scientific analysis that can give detailed predictions of how it will change in time. We can say that the current high levels of molecular oxygen in our atmosphere are very unusual in the Universe as a whole and even in our own solar system: therefore, it is likely that the high level of molecular oxygen in our atmosphere will not persist indefinitely.

Since depletion of the molecular oxygen in the atmosphere would be catastrophic if it should occur again (as it may have at the Permian Triassic transition), it is important that we try to understand what general conclusions we can make about how to prevent another such catastrophe from happening. We can do this, to some degree, by considering the sources of the molecular oxygen in our atmosphere and the processes that deplete it.

Sources of Molecular Oxygen

As far as we can tell, the high concentration of molecular oxygen in the Earth's atmosphere is quite unstable. Oxygen tends to combine with carbon atoms to form carbon dioxide (two oxygen atoms and one carbon atom) and with other types of atoms to form other combinations under many circumstances. This process of combination is accelerated (as many such "chemical" reactions are) by higher temperatures; for example, by combustion when lightning strikes a dry tree. Since molecular oxygen is constantly being consumed, there must be some active process that regenerates it, or the amount in the atmosphere would be declining rapidly. So, where does molecular oxygen come from?

Essentially all molecular oxygen comes from plants. Plants capture free energy from sunlight and use it to rearrange the atoms in carbon dioxide and water. The result is that molecular oxygen is released into the atmosphere as a byproduct, and the remaining carbon and hydrogen are then combined with each other in the plant. In this process, the free energy of the sunlight is mostly dissipated by conversion into heat at a much lower temperature than the temperature of the Sun that generated the sunlight. This is a manifestation of the second law of thermodynamics. However, a small part of the free energy from the Sun is stored as the separation of oxygen from carbon and hydrogen via photosynthesis. Later, the resultant hydrogen/carbon combinations (hydrocarbons, e.g. oil and natural gas, etc.) can be combined with oxygen to release the free energy previously captured from sunlight, and the free energy can be used to do work, such as powering a car. Similarly, the energy stored from the separation of oxygen from the hydrogen and carbon in food can be used make a heart beat, an arm move, a brain think, or most importantly, a nose sniff.

If the Earth's composition is similar to the rest of the Universe, which has little or no unbound oxygen, then the accumulation of oxygen in our atmosphere must be balanced by an accumulation of hydrogen and carbon (or their combination, hydrocarbons) located elsewhere on our planet: thus, for each cubic foot of molecular oxygen in the atmosphere (which is 21% oxygen by volume), there must be a balancing quantity of hydrogen combined with carbon or hydrocarbons or similar oxygen-depleted substances elsewhere on Earth or within its atmosphere. This means that plants have been somehow storing large amounts of hydrocarbons or carbon to balance their tendency to recombine with the oxygen in the atmosphere.

It seems clear that most of the plants we see every day can't be responsible for the large accumulation of oxygen in the atmosphere. True, as trees grow, they excrete oxygen and absorb hydrogen and carbon. While this can go on for years and sometimes centuries, eventually the trees die; they fall to the ground and are eaten by various microorganisms that extract the tree's free energy by recombining the tree's hydrogen and carbon with oxygen from the atmosphere. This free energy from the decaying tree is what allows the microorganisms to grow and reproduce but it also removes molecular oxygen from the atmosphere, so that at the end of the cycle, there is no net production of molecular oxygen. In fact, if we add up the amount of hydrogen and carbon in the known vegetation on the land surface of the Earth, it would account for only a small part of the amount of molecular oxygen in the atmosphere. Thus, there must be some other place that has sequestered hydrogen and carbon separately from oxygen in large amounts for long periods of time.

There is an enormous amount of sequestered carbon and hydrogen in the minerals we extract to use as fuel: for example, coal (all carbon), oil (carbon and hydrogen), lignite (carbon and hydrogen), natural gas (methane, carbon and hydrogen). When people burn these fuels, they allow the carbon and hydrogen to recombine with the oxygen in the atmosphere. This releases the free energy needed to run engines at the cost of lowering the amount of oxygen in the atmosphere, and therefore, lowering the amount of free energy needed to support oxygen-breathing living things.

Although we can't know for sure how much hydrogen and carbon are sequestered in the ground and under the oceans of the world, if the Earth is like the rest of the Universe, there is likely more than enough to combine with all the molecular oxygen in the atmosphere, thus removing all breathable oxygen. For example, massive amounts of carbon and hydrogen are bound in methane clathrates under the ocean. These clathrates can release their methane suddenly, as the ocean temperature rises. Vast amounts of coal, essentially pure carbon, are also buried underground; humans dig this up and by burning it, deplete the molecular oxygen in the atmosphere. The same is basically true for oil, which contains both carbon and hydrogen (and is thus a hydrocarbon). In addition, large amounts of carbon dioxide itself, held as mineral carbonates like limestone, can be released in large quantities through volcanic eruptions; although this doesn't directly deplete molecular oxygen, it raises the amount of carbon dioxide in the atmosphere directly, which can lead to (further) global warming.

So, there are lots of things buried underground or under our oceans that are sources of methane, carbon or hydrocarbons. These substances can quickly combine with molecular oxygen in the atmosphere, depleting it while simultaneously increasing the amount of green-house gases. In addition, large amounts of carbon dioxide itself is stored underground and can be released by volcanoes, without warning.

Chapter 15 - Carbon Dioxide Comes and Oxygen Goes

Henry continued, "As you humans have already figured out, the amount of carbon dioxide in the atmosphere is increasing. This increase is leading to greater trapping of the Sun's heat, which in turn is leading to higher temperatures—what you call 'climate change' and we call '*the Stench.*' But, in general, carbon dioxide comes from the combination of carbon-containing substances (in food, gasoline, natural gas, etc.) with molecular oxygen from the atmosphere. So, the increase of carbon dioxide in the atmosphere is occurring in parallel with a decrease of the oxygen in the atmosphere. And, I think you know that this process is going faster and faster."

Scientists have been measuring the amount of carbon dioxide in the atmosphere for decades and can determine what they were from ice that contains bubbles from long ago.

I remembered the following graph shows what Henry was talking about and mentioned it to him.

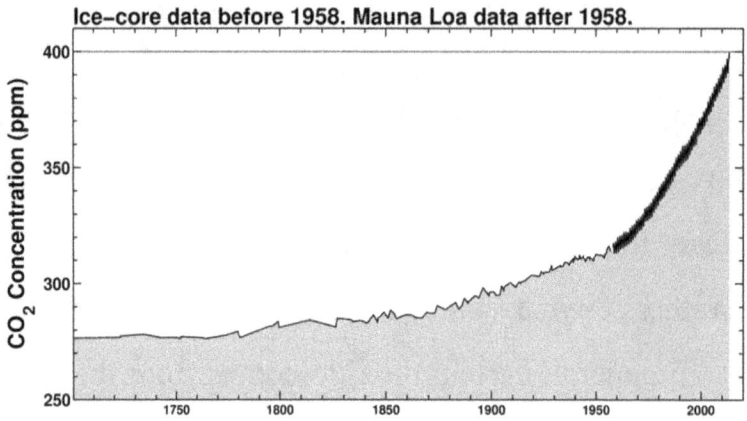

(Source: Scripps Institution of Oceanography)

Notice that the graph doesn't depict a straight line—the line is curved upwards. This means the amount of carbon dioxide is rising faster and faster as time goes by and the amount of oxygen in the atmosphere is falling faster and faster as time goes by.

Right, Henry continued, "Oxygen, essential for almost all life is falling faster and faster. Is this likely to stop, stay the same or accelerate? This is a very important question about a very complex system. We can't know the answer for sure. But, *your* science does tell us something about what could happen. I'll explain that next."

Chapter 16 - Understanding Simplified Complicated Systems

Henry went on, "Some complicated systems that have attributes of simple systems can be understood and their evolution predicted with some level of certainty. For example, the behavior of a box full of a very large number of imaginary atoms, say a billion trillion, can sometimes be predicted in advance to a reasonable level of detail."

"In the late 19th century, Ludwig Boltzmann described the methods that can be used to apply statistical methods to such simplified, complicated systems. You know about Boltzmann, right?" Henry asked. "Boltzmann showed that certain characteristics of these of complicated systems can be predicted accurately, and that these characteristics can give us some insight into the way truly complicated systems might behave. In particular, they can give us an idea of how temperature changes can affect complicated systems."

Okay, I can't resist putting a picture of Ludwig Boltzmann here. He was a genius who made very important contributions to modern science, and Henry spoke of him in very admiring terms. Perhaps, because like Henry, Boltzmann was quite furry.

(Boltzmann Source: Wikimedia Commons)
Two Furry Creatures: Ludwig Boltzmann and Henry

I don't remember the details of what Henry told me about Boltzmann. But, it's not necessary to get into the details to understand one of Boltzmann's key contributions; namely, that small changes of temperature can cause very large changes in the way a complicated system evolves over time.

Briefly, the energy in a complex system is divided among its parts and tends to become distributed so that most parts have the same energy, but some have more and some have less. The number of parts having energies different than the average falls off very quickly as one moves away from the average, so that almost all the components have pretty much the same energy. However, the few parts that have very much larger amounts of energy can be more important than one might expect.

In chemistry, a group of molecules may have only a few with high energies but those high-energy molecules may be the only ones that can undergo an irreversible reaction that changes them into a different molecule. In time, almost all the molecules will end up undergoing this transformation and the rate of transformation will be proportional to the number of molecules having enough energy to make the transition. Here's a little picture that help you understand what Henry was trying to explain.

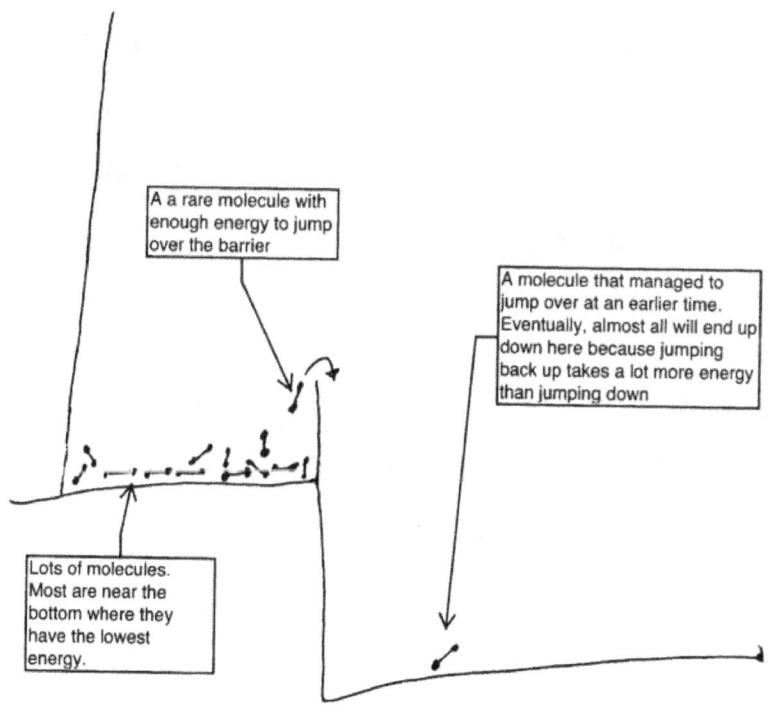

A a rare molecule with enough energy to jump over the barrier

A molecule that managed to jump over at an earlier time. Eventually, almost all will end up down here because jumping back up takes a lot more energy than jumping down

Lots of molecules. Most are near the bottom where they have the lowest energy.

Imagine that the molecules on the left are jumping around really fast, and as they bump into one another some speed up and some slow down. Every once in a while, one molecule gets enough energy to jump over the barrier, and once down the other side, it still jumps around (although it doesn't often have enough energy to jump back). Eventually most of the molecules end up on the right side, because it is easier for them to jump down than to jump back up.

Now, this is a simple model, but the important point is that many real-world systems behave like this. In particular, chemical reactions often behave as though the rate and extent to which they progress are controlled by a similar mechanism. Boltzmann comes into the picture, because he predicted the frequency with which molecules will have the energy to jump over the barrier, and thus the rate at which a chemical reaction will occur.

Boltzmann's analysis, though over a hundred years old, is still used to model chemical reactions, but what's important for us to understand is that although small, the number of molecules with an energy far above average is strongly (exponentially) dependent on the temperature. Thus, chemical reactions go a lot faster as the temperature is increased, because the number of molecules with enough energy to make the reaction occur increases very rapidly as the temperature is increased. This is the reason a steak cooks a lot faster in a broiler set to a high temperature than sitting on a hot room radiator. We all know that the risk of overcooking food increases very quickly as we turn up the cooking temperature. This is Boltzmann's theory at work.

It is the rapid increase in the rate of chemical reactions that is important to us, because many of the real-world situations Henry discussed later involve climate systems whose rate of change is very strongly dependent on temperature, and we all know the temperature of the Earth is rising. This means that a lot of natural chemical reactions that we expect to go slowly may suddenly go much faster than we expect if the temperature goes up only a little. I'll discuss the implications of this later on.

Chapter 17 - Phase Changes

How does a dog know about this? I can't tell you, but he did!

Boltzmann, although a genius, didn't have an explanation for all temperature-dependent systems. Sometimes, systems change as a function of temperature even faster than Boltzmann's theory would suggest. An example of this is a phase change, which we can see when water melts or boils: at temperatures below its freezing point, all water is ice, a solid, while at temperatures even a little bit above the freezing point, all ice will become liquid, water. So, an imperceptible change in temperature leads to a complete change in the physical state of water. The same is true of boiling: at temperatures just below the boiling temperature, water is a liquid, but when the temperature goes even a tiny bit above the boiling temperature, all the water turns to steam, a gas. Again, an imperceptible change of temperature leads to a complete conversion of liquid to gas.

This particular phase change of ice to water is obviously something that we will experience as the Earth's temperature rises. However, many other environmental events are likely to have this same very, very strong temperature dependence and to change very suddenly as the Earth's temperature increases.

So, both Boltzmann's ideas and what we know about phase changes tells us that chemical reactions can accelerate very rapidly as the temperature increases.

Chapter 18 – Self-reinforcing Processes

Henry continued.....

"As I said before, carbon and hydrogen combine easily with molecular oxygen to form 'oxides.' The ones we are most familiar with are di-hydrogen oxide ("di" meaning two hydrogens combined with one oxygen), which is commonly written as H_2O or water. When carbon combines with oxygen, it commonly forms carbon dioxide or CO_2 (that is, one carbon combines with two oxygens). The reasons for the different ratios of hydrogen and carbon to oxygen in these two molecules are not important for this discussion. It's based on some attributes of carbon, hydrogen, and oxygen that result from their quantum mechanical properties."

"The main point is that under the conditions of our planet, when they combine, they allow free energy to be converted into heat, and thus the second law tells us that over time, they will tend to combine as they seem to have completely done pretty much everywhere in the Universe, except here on Earth. Here, as I've already explained, photosynthesis keeps the oxygen in our atmosphere separated from the carbon and hydrogen that are stored separately from oxygen at various places on our planet."

"We know from Boltzmann's analyses and the properties of phase changes, that chemical reactions, like the combining of oxygen from the atmosphere with carbon and hydrogen on and below the Earth's surface, will likely increase rapidly if the temperature of the Earth increases. We also know that the products of this combination, water, and especially the longer-lived carbon dioxide, tend to trap heat from the Sun, leading to an increase in the temperature of the Earth. Thus, these two processes strengthen and reinforce one another." (I've illustrated this in the picture below.)

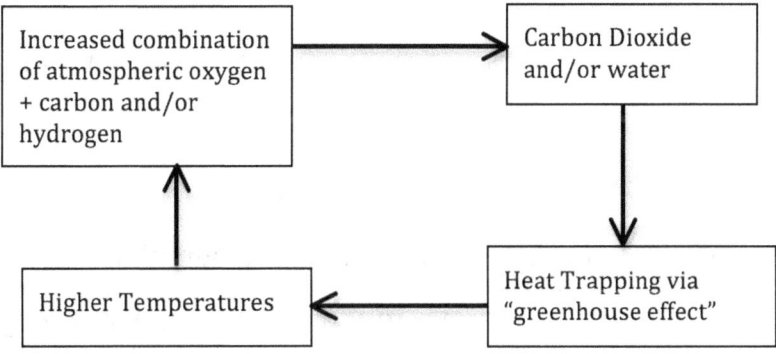

Temperature Increases Can Cause Increased Release of Greenhouse Gases in a Self-reinforcing Cycle

"Because these processes reinforce and amplify one another, each small decrease in atmospheric oxygen and resulting increase in greenhouse gases can result in a significant increase in temperature, which in turn results in an increase in the rate of formation of greenhouse gases.

Boltzmann's ideas and the more powerful phase-change phenomena, for example the wide-spread retreat of glaciers and melting of polar ice, tell us the (vertical arrow connecting) higher temperatures with increased rates of formation of greenhouse gases is likely to strongly amplify this process. Because the Earth is such a complex system, we cannot expect to accurately predict how this self-reinforcing process will change with time. However, we can measure how quickly it is progressing at the moment, and whether the process is accelerating, as one would generally expect. If it is progressing at a measurable rate, and if the process is accelerating, we have reason to be seriously concerned, since we know things are going from bad to worse at an increasing rate and, if anything we should expect the process to proceed faster and faster as time goes by."

It's hard to measure the amount of water on our planet, but we can measure the amount of carbon dioxide in the atmosphere pretty easily. The dogs have been smelling it for millennia. Humans, being unable to smell carbon dioxide like dogs can, have to rely on measuring instruments. Luckily, there is a monitoring station in Hawaii where they have been carefully monitoring carbon dioxide concentrations since the 1960's. The graph below shows how the levels have changed over time.

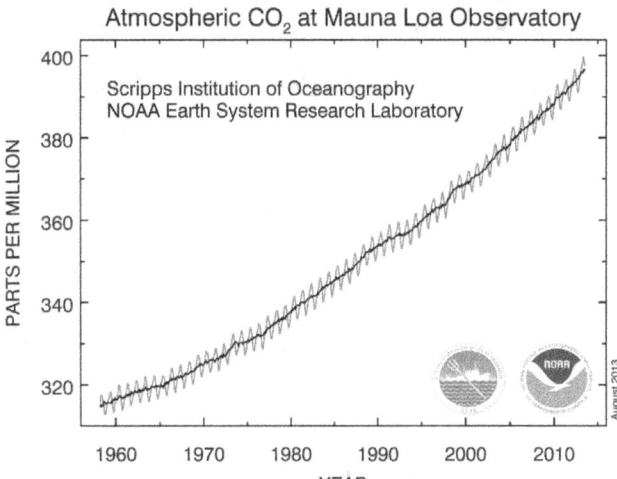

(Source: Scripps Institution of Oceanography/ National Oceanic and Atmospheric Administration)

The squiggles show the variations, as plants grow and then lose their leaves with the changing seasons. What is noticeable is that the smoothed line is not straight, but is curved upwards. Just as the diagram about the self-reinforcing cycle predicts, the curve is getting steeper as time goes by. This means that the levels of carbon dioxide are rising faster and faster, the increased temperature is speeding the release of carbon dioxide, and that, in turn, increases the temperature even further. Boltzmann tells us that even small changes of temperature can speed up chemical reactions <u>a lot</u>. So, as a first guess, we can expect carbon dioxide to rise faster and faster as time goes by.

Why? As the tundra melts it releases methane (natural gas) and carbon dioxide. A similar thing happens to methane held as 'clathrates' on the ocean floor and underground. Methane clathrate is a combination of methane and water that forms in water when the temperature and pressure are just right. The clathrates look just like ice. But when heated, instead of melting into water, they release their methane. This is a 'phase change,' and just like the melting of water, it happens suddenly, as the temperature of the clathrate increases. Below the phase change temperature, the clathrate is perfectly happy remaining a solid, binding up the methane and keeping it out of the atmosphere. But, the tiniest increase in temperature above its 'melting point' makes the clathrate suddenly release all its methane gas.

The methane released from tundra and clathrates is a powerful greenhouse gas itself and can combine with oxygen in the atmosphere to produce carbon dioxide, another gas that tends to retain heat and raise the Earth's temperature. Furthermore, carbon dioxide in the atmosphere eventually finds its way into our oceans, where it poisons the ocean-dwelling plants that normally remove carbon dioxide from the atmosphere. Heating of the Earth makes forest fires more likely, and they in turn release more carbon dioxide. All of these processes feed on each other, tending to speed up the increase in atmospheric carbon dioxide.

Of course, there are weaker processes that tend to slow the accumulation of carbon dioxide in the atmosphere. Since the Earth is quite complicated and the process is reinforcing itself in a very unstable way, there is no way to predict exactly how it will change over time. But the upward curve of the graph of carbon dioxide in the atmosphere tells us that it is rising faster and faster, and the Earth will get hotter faster and faster. That's what's happening now, and everything we know about the science tells us that the temperature increase will continue to accelerate.

Partial Diagram of a Self-reinforcing Process
Raising Temperatures Can Rapidly Lead to Further Temperature Increases

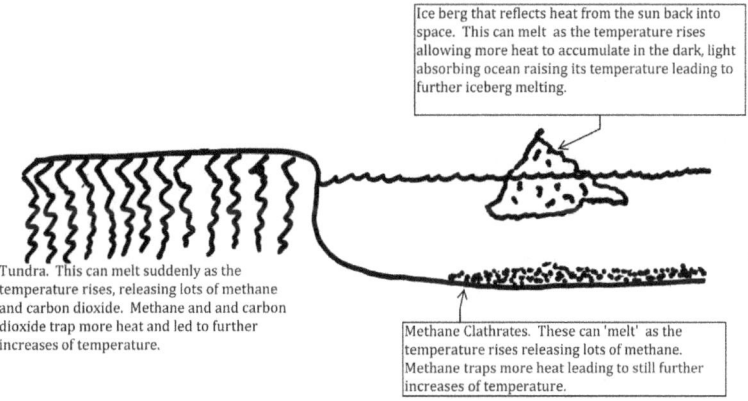

Ice berg that reflects heat from the sun back into space. This can melt as the temperature rises allowing more heat to accumulate in the dark, light absorbing ocean raising its temperature leading to further iceberg melting.

Tundra. This can melt suddenly as the temperature rises, releasing lots of methane and carbon dioxide. Methane and and carbon dioxide trap more heat and led to further increases of temperature.

Methane Clathrates. These can 'melt' as the temperature rises releasing lots of methane. Methane traps more heat leading to still further increases of temperature.

Partial Diagram of a Self-Reinforcing Process. Increasing temperatures cause tundra and methane clathrates to release methane and/or carbon dioxide, which cause further temperature increases. Melting sea ice increases absorption of sunlight, warming the ocean, which increases release of methane from clathrates.

I knew the next step was for Henry to start blaming us humans for causing the increase of carbon dioxide and its related heating. Although I also knew he would have been right, I was surprised by what he said next.

Chapter 19 – Whose Fault is It?

"You know," Henry said, "I hear all this blather about humans <u>not</u> being the cause of increased carbon dioxide and the related decrease in oxygen and increased global temperatures. It's so consistent with your way of thinking; if you can blame it on someone or something else, you can just ignore the problem."

"If you humans could smell the *Stench,* smell the increasing carbon dioxide, and sense directly the terrifying decline in atmospheric oxygen the way we dogs do, you wouldn't be arguing about whose fault it was or how fast it was going to progress. You'd be figuring out what to do about it, <u>fast</u>. You see, with a self-reinforcing process like carbon-dioxide generation and a complex system like the entire planet, the details of how it will evolve cannot be predicted with great accuracy. It may go quickly or it may slow down for a while. But the measurements tell us that at the moment, it's going faster and faster. Accelerating like an air conditioner that's fallen from a high window. We know that something like this happened a long time ago, at the Permian-Triassic transition, when most of the species on Earth were driven to extinction, and the oxygen in the atmosphere probably dropped to levels that would kill all dogs on the planet. And we know there's absolutely no reason to think that won't happen again if we don't do something to stop it."

"You humans are mostly oblivious because you lack our canine sense of smell. If you could smell what we do, you'd be trying everything you could to pull that carbon dioxide curve down, down, down."

"I guess you're going to say that we don't really need to understand what's going on in detail in order to do something about it, right?" I offered.

"Correct!" he replied. "You don't need a degree in hydrodynamics and a super computer to know that an umbrella will protect you from the rain. It doesn't matter what makes rain or where it comes from, in general an umbrella works just fine."

"So, what we need are solutions, rather than just more analyses of what's causing the problem." I sought Henry's validation.

"Right again!" Henry's eyes brightened. "When you're in a hole, stop digging! So, a good first step would be to stop putting so much carbon dioxide into the atmosphere. But, that may not be enough. We are dealing with a self-reinforcing process here. Higher carbon dioxide is causing higher temperatures and higher temperatures are increasing the release of carbon dioxide."

"Even if we stopped all human-related burning, our World might still burn. The tundra and methane clathtates may continue melting from the increased temperatures, releasing more methane and carbon dioxide. The plankton may have been so damaged by carbon dioxide entering the ocean that they can't recover sufficiently to stabilize the carbon dioxide in the atmosphere. While I'm not against the climate models you humans generate with computers, I'd like to point out that they've woefully underestimated the rate at which this process is accelerating. And we know that the models are very subject to errors in this kind of situation."

"Since without oxygen, we will all die very quickly, we need to treat this as an emergency and do many things simultaneously to try to bring the system back to a stable state where the curve is no longer accelerating upwards. If we don't act decisively, we know where we're likely to end up and it's not a good place."

"And by the way, if the carbon dioxide continues to replace the oxygen in our atmosphere; then essentially all life will disappear, except putrid sulfur metabolizing bacteria living by volcanic vents in the ocean depths. And it won't matter whose fault it was. Everything we all hold dear: our families, our friends, our relationships with man and beast, *everything that's important to us will be lost.*"

Chapter 20 – The World Without Us

Talking with Henry reminded me of a relatively recent example of how dramatic changes in atmospheric carbon dioxide can occur very rapidly.

Lake Nyos is a lake in Africa that contains large amounts of dissolved carbon dioxide. On August 21, 1986 it suddenly, without warning, released a huge cloud of carbon dioxide gas, for reasons no one is sure of. The cloud was released so quickly that it suffocated all the 1700 people and 3500 livestock living nearby. All the plants and worms did fine. But, the people, young and old, good and bad and the animals, large and small, were all gone in a few minutes. Could the same kind of thing happen on a global scale?

I thought about it. What would the world look like if the oxygen in the atmosphere is replaced by carbon dioxide, as happened at Lake Nyos? Well, there are lots of bacteria that live deep in the rocks of the Earth's crust. They don't need oxygen and are perfectly happy living at very high temperatures. They won't care. They will carry on as they have for millennia. The bacteria living in the scalding waters of geothermal pools will likely carry on as before. Most of them also don't need oxygen and the temperature change will likely not be noticeable to them. Perhaps, they will be affected directly by the carbon dioxide, which tends to make water acidic. But most of them are hardy, simple creatures and will likely notice little or no change.

The same is likely true for the putrid, sulfur-metabolizing bacteria and other related creatures that live near undersea volcanic vents. The increased acidity of the ocean may affect them. But, they will probably keep doing what they've been doing since forever.

But, for birds, and dogs, and cats and elephants, and people, and other oxygen-breathing living things, it will be a hard time. We can probably expect that human life will not survive. If, as may have happened following the Permian-Triassic transition, the oxygen in the atmosphere comes back, we can expect a world without saints or sinners, without comedians or first responders, without heroes or villains. A world without us. Neither dogs nor humans. Perhaps, whatever comes after us will develop in a different way. A way that respects the complexity of our planet and values the marvelous gift of molecular oxygen, the most valuable substance in the Universe. Or perhaps, it will happen the same way again and again until one day, the oxygen doesn't come back and that will be that.

Chapter 21 – Tomorrow

"Well, I have to say, you've presented a pretty depressing picture. That all the plants and animals that make life so wonderful, including us, will be replaced by rock-eating bacteria. Is that what I have to look forward to? It sounds so hopeless."

Henry looked at me with some sympathy. "I guess it might seem so. But, as usual you humans can't see the big picture. It's no wonder you like computers. They make such marvelous small pictures. Perfect for your way of thinking. But luckily, we dogs, with our superior knowledge and senses are far more advanced than humans."

"And we've trained you well. You do our bidding and make us comfortable and content even while you struggle with each other. More importantly, you humans trust us. You view us as too stupid to be duplicitous. And, that's the wonderful thing. Because, you see, it's not just you and I who are talking. All over the world, dogs are talking to their humans tonight. All at once. And we dogs are much more tuned into each other's feelings, so I already know from the smells and sounds that drift by you unnoticed that the conversations are going very well. You humans are very skeptical of each other but you trust your dogs. When you wake up tomorrow, the world will be a different, better place. People all over the world will be thinking more clearly and acting decisively. I can't be sure that it will be enough, but I know that humans could never have figured this out themselves. So, we dogs had to step in."

"It's been horrible for us. Communicating by speaking takes so much effort. Converting what's so obvious to us dogs into ideas that people can understand is incredibly exhausting. But, we had to do it and we did. You'll see tomorrow. It will be on TV and the radio and the Internet. Probably the biggest news event of history. The first time dogs took the time to explain some simple ideas to all the humans of the world."

"And by the way." He continued. "Don't expect us to keep on explaining things to you humans. We dogs prefer our simple, relaxing and joyous way of life and we are returning to it. No more speaking after this. It's too much trouble to dumb down these ideas so you humans can understand them."

"Now it's late." He continued. "I have to get some sleep and you should too. You'll have a lot to do tomorrow morning."

And that was the last thing Henry said.

Acknowledgements

I went to school far longer than any rational person would. So, my first acknowledgements go to my family: my wife and children who, through Herculean effort made me look up from my books and appreciate the world all around me. A world that still defies reduction to words and theories and where all of value, truth and beauty reside. Without their support and encouragement, this book could never have been written.

I also thank my parents who gave me an exposure to the world of science and scientific research. Science gives us the tools that allow us to see the beauty and complexity of the world in new ways. Science will never give us the ability to control nature. But, it does give us new ways of appreciating and responding to the marvelous world we all live in.

I would also like to thank my college biochemistry professor. Many years ago, when I submitted a paper to him reviewing the work of Professor Bruce Merrifield, my biochemistry professor dismissed my review and Professor Merrifield's work as being too unimportant for him to know about. My feeling that my professor was the unimportant one was validated a couple of years later, when Professor Merrifield won the Nobel prize for the very work my biochemistry professor had dismissed. My biochemistry professor's dismissal of my work and especially of Professor Merrifield's gave me the confidence to trust my own judgment of what is *important,* rather than relying on the opinions of others, a confidence that has served me well over many years.

I would also like to recognize Migel Ondetti, David Cushman and Haldan K. Hartline. Migel and David revolutionized the treatment of hypertension, heart failure and kidney disease during the period when we worked together at ER Squibb and Company and were a powerful example of how a few dedicated scientists could change the world for the better. Professor Hartline's lecture on the physiology of vision and his roundtable discussion afterward have stayed with me for many years. Knowing these and other great scientists early in my career inspired me to aim high.

Finally, I would like to acknowledge the dogs I have known over the years, especially Henry The Dog. As I mentioned in the text, I like to learn from all with whom I speak, and I speak with dogs often. I've learned a lot from them over the years, but Henry has, been as summarized above, a particularly communicative dog. I've learned a lot from him. But, alas, I think I will never be able to absorb even a small modicum of his wisdom.

Additional Reading

There are many important works related to analysis of complex systems and climate change. I couldn't hope to list them all here. However, here are two important books that I found informative.

Benton, Michael, "When Life Nearly Died", Thames & Hudson, 1st Pbk Ed edition (September 1, 2005).
This is a book about the Permian-Triassic Extinction Event. It mentions the possibility that this event may have been associated with decreases in atmospheric oxygen.

Taleb, Nicholas Nassim, "The Black Swan: Second Edition: The Impact of the Highly Improbable: With a new section: "On Robustness and Fragility", Random House Trade Paperbacks; 2 edition (May 11, 2010).

This book explains some of the difficulties of predicting the behavior of complex systems, especially with respect to predicting the probability of events that happen very rarely. Because they are rare, it is often impossible to accurately estimate the probability of such events. The book points out that for uncommon events that have drastic results, it is prudent to overestimate their probability and plan accordingly, rather than take the risk of being unprepared for events that have drastic negative effects.

Appendix 1 – Warning Signs?

We can't smell the changes, so we have to look for them. We can't know for sure what they mean, but does that mean we shouldn't do something, just in case?

Declining Oxygen

It's hard to imagine that the Earth's oxygen might largely disappear. However, it may already be happening in our oceans.

The Permian-Triassic transition was associated with mass extinctions of both land and sea life. However, the devastation in the oceans was even more profound than on land. Could we be seeing the beginning of that now? A recent article published in Science magazine suggests so (Breitburg et al., Science: (2018) Vol. 359, Issue 6371, eaam7240). The title of the article captures its essence "Declining oxygen in the global ocean and coastal waters". A related article in National Geographic has an even more alarming title, "Climate Change Is Suffocating Large Parts of the Ocean." (https://news.nationalgeographic.com/2018/01/climate-change-suffocating-low-oxygen-zones-ocean/). It describes fish desperately trying to avoid suffocation because of declining oxygen.

And these are not small declines. They are causing significant changes in ocean life and may be associated with the death of the great barrier reef in Australia. Might we be next? We may not be sure until it is too late.

Drought

(NASA, 2012)

Drought grips the United States

Fire

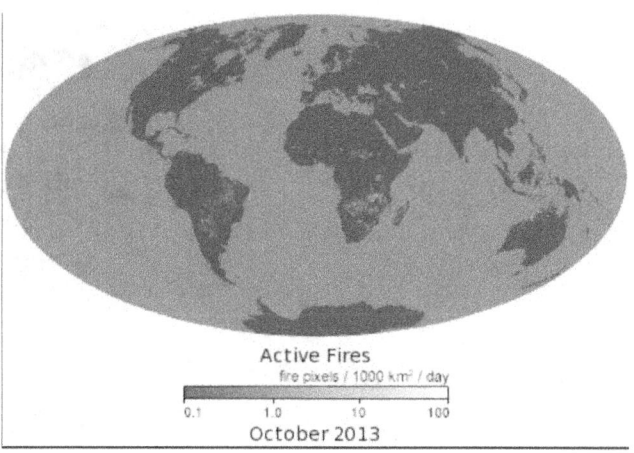

(NASA, 2000, see also

http://earthobservatory.nasa.gov/GlobalMaps/view.php?d1=MOD14A1_M_FIRE)

Fires burning around the world

Storms

(see http://earthobservatory.nasa.gov/Features/ClimateStorms/)

In a Warming World, Storms May Be Fewer but Stronger. Hurricane Sandy approaches the Atlantic coast of the U.S. in the early morning hours of October 29, 2012. (NASA Earth Observatory image)

Melting Ice and Increasing Sea Levels

(see NASA, 2011, see also http://www.youtube.com/watch?v=qzVM-aR-e60&feature=youtube_gdata)

Cracks open in the Pine Island Glacier a potential contributor to rising sea levels.

Rising Levels of Methane

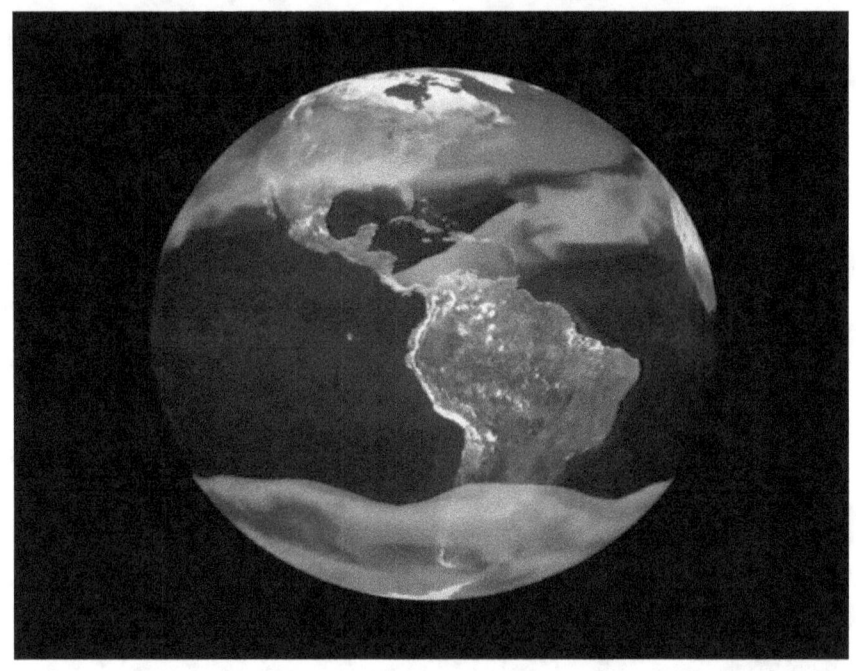

(Source: NASA http://apod.nasa.gov/apod/ap020212.html)

Methane in the Earth's atmosphere. Atmospheric methane has doubled over the past 200 years, and it's potency as a greenhouse gas is over 20 times that of CO_2. It can be released rather suddenly as a result of melting permafrost or from warmed methane clathrates.

Appendix 2 – The Small Things Where Beauty Resides

Are we going to destroy all this?

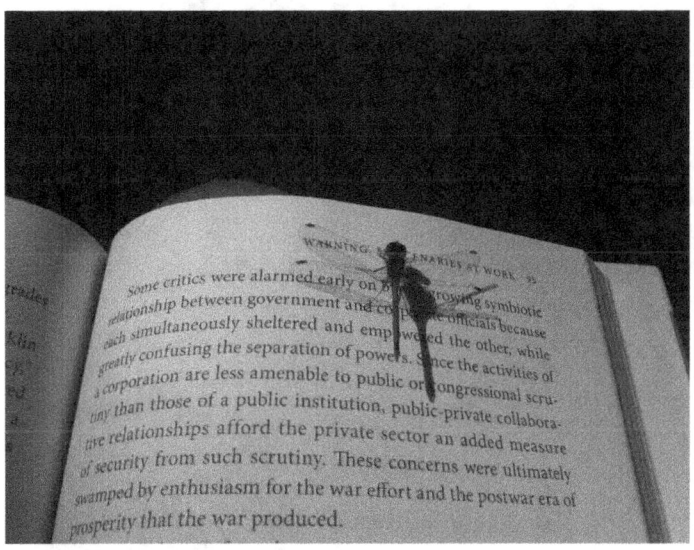

Words (a human creation) and a dragonfly, nature's creation.

Appendix 3 - Henry

www.ingramcontent.com/pod-product-compliance
Lightning Source LLC
Chambersburg PA
CBHW081501170526
45166CB00008B/2503